Lake Manapouri, Fiordland National Park.

Introduction

New Zealand is rich in natural wonders, with diverse landscapes, numerous pristine wilderness areas and intriguing, unique native flora and fauna.

For many thousands of years these remote South Pacific islands lay undiscovered by humans, their young mountains, forests and wildlife left to evolve in splendid isolation. As one of the last land masses of its size to be populated, human impact on the environment has been relatively recent, with the first ancestors of Maori arriving from eastern Polynesia sometime around 800 to 1000 years ago. Although cities and towns, farms, orchards and vineyards now cover much of the land, large areas of natural wilderness remain.

Today over one third of New Zealand's area is protected for its conservation values. From subtropical to subantarctic climes, these areas encompass a wide range of wilderness: dense rainforests to subalpine tussock lands; active volcanoes to the Southern Alps with their snowfields and glaciers; wide river valleys and deep glacial lakes to shallow estuaries and golden beaches.

Long isolation and lack of mammalian predators and browsing animals meant New Zealand's plants and animals evolved in unique, often startling ways, and that a high proportion of them are endemic. These species and their habitats are protected in a network of national parks, forest parks, reserves and sanctuaries. Importantly, through a host of nature tourism ventures and excellent visitor facilities, everyone can visit and appreciate New Zealand's natural wonders.

Fiordland National Park

World Heritage-listed Fiordland is a vast wilderness of mountains, fiord-indented coastline, lakes, rivers and forests. Glaciation has marked the landscape, leaving features such as alpine tarns at Key Summit (above), and mountains that rise sheer from glacier-gouged fiords, like Mitre Peak from Milford Sound/Piopiotahi (right). In the Hollyford Valley (overleaf) lichen-covered forest is a consequence of the park's constant and heavy rainfall.

Queenstown and the Southern Lakes

On the drier, eastern side of the South Island's main divide, the bed of Lake Wakatipu (above), New Zealand's third-largest lake, has been gouged by glacial action, the scouring ice leaving its telling marks on the flanks of The Remarkables alongside. Introduced, autumn-flowering trees at Arthurs Point, near Queenstown (left), add colour to the landscape's natural brown and golden hues. However, in nearby Skippers Canyon (right), native tussock dominates.

Coastal Otago

Otago's coastline supports a remarkable diversity of wildlife. In the remote, southern Catlins region sea lions, fur seals, elephant seals, Hector's dolphins, fish and a host of seabirds and shore birds are attracted to habitats such as Purakaunui estuary (left) and the productive feeding grounds off Tokata, or Nugget Point (below right). Colonies of yellow-eyed penguins, little blue penguins, spotted shags, sooty shearwaters, Australasian gannets and royal spoonbills occupy the beaches, headlands and rock stacks around the point. A network of small, scientific reserves and a forest park protects the special habitats of the Catlins, and local residents have developed excellent nature tourism ventures so that they can share their natural heritage with appreciative visitors. On the northern Otago coast large, curious-looking marble-shaped rocks, known as the Moeraki Boulders (top right), lie strewn along the beach. Depending on the explanation you prefer, these are 65 million-year-old concretions, or gourds from an ancient migratory canoe.

Central Otago

Central Otago's landscape inspires artists and poets, with its sprawling plains and ranges, tussock and rocky tors, glacial lakes, bold colours and clear, sharp light. Introduced poplar trees add autumnal hues to the Lake Wanaka shore (right). At St Bathans, the intense colouring of the Blue Lake (left) comes from minerals in the surrounding cliffs. Stark browns greet travellers on tussock-covered Lindis Pass (below) that crosses from Otago to Canterbury.

Canterbury

Ancient weathered rock has created many dramatic and curious landforms throughout Canterbury. At Gore Bay, on the east coast, the eroded siltstone pinnacles known as The Cathedrals (left) can be seen from either the coastal road or a short walking track. Inland at Castle Hill (top right), limestone canyons, archways and other strange rock shapes attract both curious onlookers and serious rock climbers. Cave Stream (bottom right) is a 360 metre, twisted tunnel, eroded by water filtering through the limestone rock, which can be safely negotiated in the company of an experienced guide.

Overleaf: Mt Hutt Range and Rakaia River. The Rakaia is one of New Zealand's largest braided river systems, home to threatened bird species the wrybill plover, banded dotterel and black-footed tern. It is also an extremely popular fishery, in particular for introduced salmon. The river meanders across the Canterbury Plains, arid farmland largely sheltered from moist westerly airflows by the Southern Alps.

Aoraki/Mt Cook National Park

Mountains and glaciers are the showpieces in this World Heritage park. At 3754 metres, Aoraki/Mt Cook (right) is New Zealand's highest mountain, a drawcard for both climbers and less physically inclined tourists. In the foreground giant mountain buttercup presents its own summer show. At the base of Tasman Glacier, Blue Lake (above) is a good, albeit cold, swimming spot. At Lake Tekapo, to the south of the national park, flowering lupins (left) are in fact introduced weeds, but are forgiven by many for their colourful show.

Overleaf: Three major glaciers feed into Lake Pukaki, their finely ground glacial debris suspended in the water helping to give it an intense blue colour.

Mt Aspiring National Park

A wilderness of mountains, glaciers, snowfields, valleys and forests make up this World Heritage park, which straddles the main divide. In the west, where rainfall is heavier, rainforests are filled with luxuriant ferns (above left). The aptly named Blue River (below left) is typical of the cold, clear rivers that flow from this vast, pristine wilderness. Many flow into major valleys, such as the Haast (above).

Westland/Tai Poutini National Park

In this spectacular World Heritage park, glacier meets rainforest and wilderness extends from mountains to lagoons, lakes and the sea. Two major glaciers, Fox/Te Moeka o Tuawe (right) and Franz Josef/Ka Roimata o Hine Hukatere (below left and overleaf), are among the most accessible in the world. The glacier-formed Lake Matheson (above left) reflects the country's two highest mountains, Aoraki/Mt Cook and Mt Tasman.

Arthur's Pass National Park

This mountain park is traversed by the highest road and rail crossing in the alps, and is thus easily accessible. The eastern, drier regions are characterised by beech forests and wide, braided river valleys such as the Waimakariri (left). The walk to the Bridal Veil Falls (above right) is one of several short walks in the park. Park visitors are likely to meet kea, New Zealand's inquisitive and often precocious alpine parrots (right).

Paparoa National Park

The wild coastline and dramatic limestone formations of Paparoa are popular photo stops for those touring the South Island's west coast. Cliffs and caverns along Truman Bay (above left) typify the Paparoa coast, which is nowhere more spectacular than at the Punakaiki Pancake Rocks and blowhole (left). In the subtropical rainforest that thrives along this coast nikau palms (right) grow near the southern limit of their distribution.

Kahurangi National Park

Kahurangi is New Zealand's second largest national park, known for its marble karst landforms and vast cave systems. The park extends from alpine tops and broad, tussock-covered tablelands through valleys, lakes and forests to remote coastline, where dense, primeval forests grow right to the water's edge. The Heaphy Track, a popular 4–5 day walk in the park, crosses the northern tablelands then descends to Scotts Beach (left and below right), on the west coast. For those preferring a shorter walk, the beach, with its windswept coastal views, interesting landforms and hardy coastal forest, can be explored via a 2 hour return walk from the road end. Near here is the Oparara River (above right), which leads to the Oparara Arches, impressive limestone landforms that are accessed via a walk through luxuriant west coast forests filled with tree ferns.

Abel Tasman National Park

The South Island's north-western corner is blessed with golden beaches and intriguing landforms. In Abel Tasman National Park Totaranui Beach (below), Te Pukatea Bay (left) and Anapai Bay (overleaf) are just three of many forest-fringed beaches explored by kayakers and hikers. In Golden Bay the Archway Islands on Wharariki Beach (above right) have survived years of wave erosion and today provide a home for seals, penguins and other seabirds.

Kaikoura

There are few places in the world where people can see such a diversity of marine life, and in such a splendid natural setting, as at Kaikoura, on the South Island's eastern coast. The meeting of warm, subtropical sea currents with cold, nutrient-rich currents from the Southern Ocean produces a fertile feeding ground for marine life. Sperm whales feed close to shore, while humpbacks and orca pass by during their seasonal migrations. Dolphins, seals and ocean-going seabirds are similarly attracted to the feast, and the town's name literally means 'meal of crayfish' (or lobster). Rising close to the coastline the Seaward Kaikoura Range (top and right) adds a dramatic backdrop to the busy tourist and fishing centre. Along with the plethora of whale-watching tours, dolphin encounters and pelagic seabird cruises, the surf beaches, walking tracks and challenging climbing routes in the region add even more to the nature and adventure tourism options available.

Marlborough Sounds

Conservation and tourism are happy partners in the Sounds, a tangled maze of waterways, inlets and islands fringed with luxuriant bush and, at Christmas time, studded with crimson-blossomed pohutukawa. The Sounds provide sanctuary to a diversity of seabirds, forest birds and marine mammals. Dolphin-watch cruises, kayaking and walking or biking the Queen Charlotte Track are popular activities in Queen Charlotte Sound (these pages).

Coastal Wairarapa

A sheltered lagoon, fossil-rich limestone reef, sand dunes and sea cliffs make a spectacular combination at Castlepoint, or Rangiwhakaoma (above left and right).

New Zealand fur seals (below left) were nearly hunted to extinction for their skins in the 1700s. The species has recovered and now thrives around much of New Zealand's coastline. A huge colony lives beside the coastal road at Palliser Bay.

Te Urewera National Park

New Zealand's fourth largest national park is a vast and jumbled expanse of forest-covered ranges and valleys, lakes and rivers. In the heart of the park Lake Waikaremoana (above right) is popular for water-based recreation and the two- or three-day Lake Waikaremoana Track, which circles two-thirds of the lake. Short walks feature dense rainforest and waterfalls such as the Bridal Veil (left) and Upper Aniwaniwa Falls (right).

Hawke's Bay

The world's largest mainland gannet population lives in a dramatic setting at Cape Kidnappers (left), where several colonies live a seemingly precarious existence on windswept cliffs, rocky islets and reefs. Nature tours allow close-up views of the nesting and bonding Australasian gannets (above) and their chicks. Another dramatic Hawke's Bay landform is Te Mata Peak (overleaf), a fine viewpoint overlooking the Bay's prolific vineyards.

Taranaki

Standing sentinel on the North Island's western tip is Mt Taranaki (left), the quite symmetrical volcano that is the centrepiece of Taranaki-Egmont National Park. Also in the park, Dawson Falls (below right) tumble over lava rock. On the northern Taranaki coast sea cliffs and rock stacks such as the Three Sisters (above right) make a dramatic spectacle. The coastal Miocene fossil sequence in this region attracts geologists from around the world.

Lake Taupo and Central Plateau

This tranquil view of Lake Taupo (above) belies its fiery origins. The lake fills a massive crater blasted in one of history's biggest volcanic eruptions, when pumice, ash and lava poured over the region. Lake Taupo drains into the Waikato River; the huge volume of water restricted to a narrow chasm (left) just above Huka Falls. Volcanic mountains are also the central focus of Tongariro National Park (overleaf), a World Heritage site.

Volcanic activity

Tongariro, Taupo and Rotorua all lie at the southernmost end of the 'Pacific Ring of Fire', a volatile earthquake, geothermal and volcanic zone that marks the edge of the Pacific crustal plate. In Rotorua, tourists have long been drawn to geothermal spectacles such as Pohutu Geyser (right) and Devils Home Crater, Wai-o-tapu (below left). Tours to the active White Island volcano (above) on the Bay of Plenty coast are also popular.

Coromandel Peninsula

Forest-fringed beaches, sculpted sea cliffs and caves and island sanctuaries draw hundreds of holidaymakers to the Coromandel Peninsula each year. Particularly popular are Te Whanganui A Hei Marine Reserve and Cathedral Cove (above and above right), and Cooks Beach, with its surrounding bays and beaches, such as Lonely Bay (below right).

Waikato

Known today for its farming (left) and horse-racing industries, Waikato nevertheless contains significant natural areas protected in forest parks and reserves. An outstanding feature here is the Waitomo Caves, with their star-like glow-worms, striking limestone formations such as in Aranui Cave (above) and underground rivers. Guided cave tours range from sedate walks and cruises to adrenalin-pumping rafting and abseiling. Limestone features above the ground include this forested limestone gorge in Ruakuri Reserve (right).

Auckland's west coast

The forest-covered Waitakere Ranges form a buffer between Auckland city and the wild west coast. Barely an hour's drive from downtown visitors can be exploring windswept beaches, wetlands and rocky headlands. Two of the most popular beaches are Bethells, or Te Henga (left) and Piha (right), with its imposing Lion Rock. Most popular with a growing colony of Australasian gannets, however, are the windswept clifftops at Muriwai (above).

Northland

Dramatic seascapes, outstanding marine life and New Zealand's biggest trees are all natural features of Northland. Giant kauri trees thrive in the northern rainforest, with one of the largest, Tane Mahuta (right), reaching over 50 metres tall and estimated to be 1500 years old. Dolphins and migratory whales are attracted to the sheltered Bay of Islands, where the 'Hole in the Rock' of Piercy Island (left) and beaches such as Cable Bay on Urupukapuka Island (above) attract many visitors. Near New Zealand's northernmost tip, Cape Reinga lighthouse stands guard over the meeting of the Pacific Ocean and Tasman Sea (overleaf).

First published in 2006 by New Holland Publishers (NZ) Ltd
Auckland • Sydney • London • Cape Town

www.newhollandpublishers.co.nz

218 Lake Road, Northcote, Auckland 0627, New Zealand
Unit 1, 66 Gibbes Street, Chatswood, NSW 2067, Australia
86–88 Edgware Road, London W2 2EA, United Kingdom
80 McKenzie Street, Cape Town 8001, South Africa

ISBN: 978 1 86966 137 3

Publishing manager: Matt Turner
Editor: Brian O'Flaherty
Design: Graeme Leather

A catalogue record for this book is available from the National Library of New Zealand

10 9 8 7 6 5 4 3

Colour reproduction by SC (Sang Choy) International Pte Ltd, Singapore
Printed in China through Colorcraft Ltd., Hong Kong

Front cover photograph: Stewarts Creek Falls, Haast Pass, Mt Aspiring National Park.

Back cover photographs, from top: Mt Hutt Range and Rakaia River; Queen Charlotte Sound; Devil's Home Crater, Wai-o-tapu.